Christening the Dancer

John Amen

UCCELLI PRESS

Published by Uccelli Press
P.O. Box 85394
Seattle, Washington 98145-1394
editor@uccellipress.com
www.uccellipress.com

Founded in 2001, Uccelli Press is dedicated to publishing a select assortment of poetry, fiction, nonfiction, chapbooks, anthologies, and audio recordings.

Published in the United States of America
ISBN 0-9723231-0-4
First Uccelli Press Printing, 2003
Library of Congress Control Number: 2002114729

Cover Painting "Christening the Dancer"
by John Amen

Book Design: Mary Powers at Symbiotic Design, Charlotte, NC

Photos: Ginger Wagoner at PhotoSynthesis, Inc.

Marketing and Public Relations: Marketing Matters

For more information please contact
marketingmatters@carolina.rr.com

Christening the Dancer
John Amen

"I know everything but myself."

—François Villon, "Ballade"

"I praise what is truly alive,
what longs to be burned to death."

—Johann Wolfgang von Goethe, "The Holy Longing"

"I can't help myself: the cathedrals burst with organ-fugues.
I want to make a new sun."

—Hugo Ball, "The Sun"

Part One

Part Two

Part Three

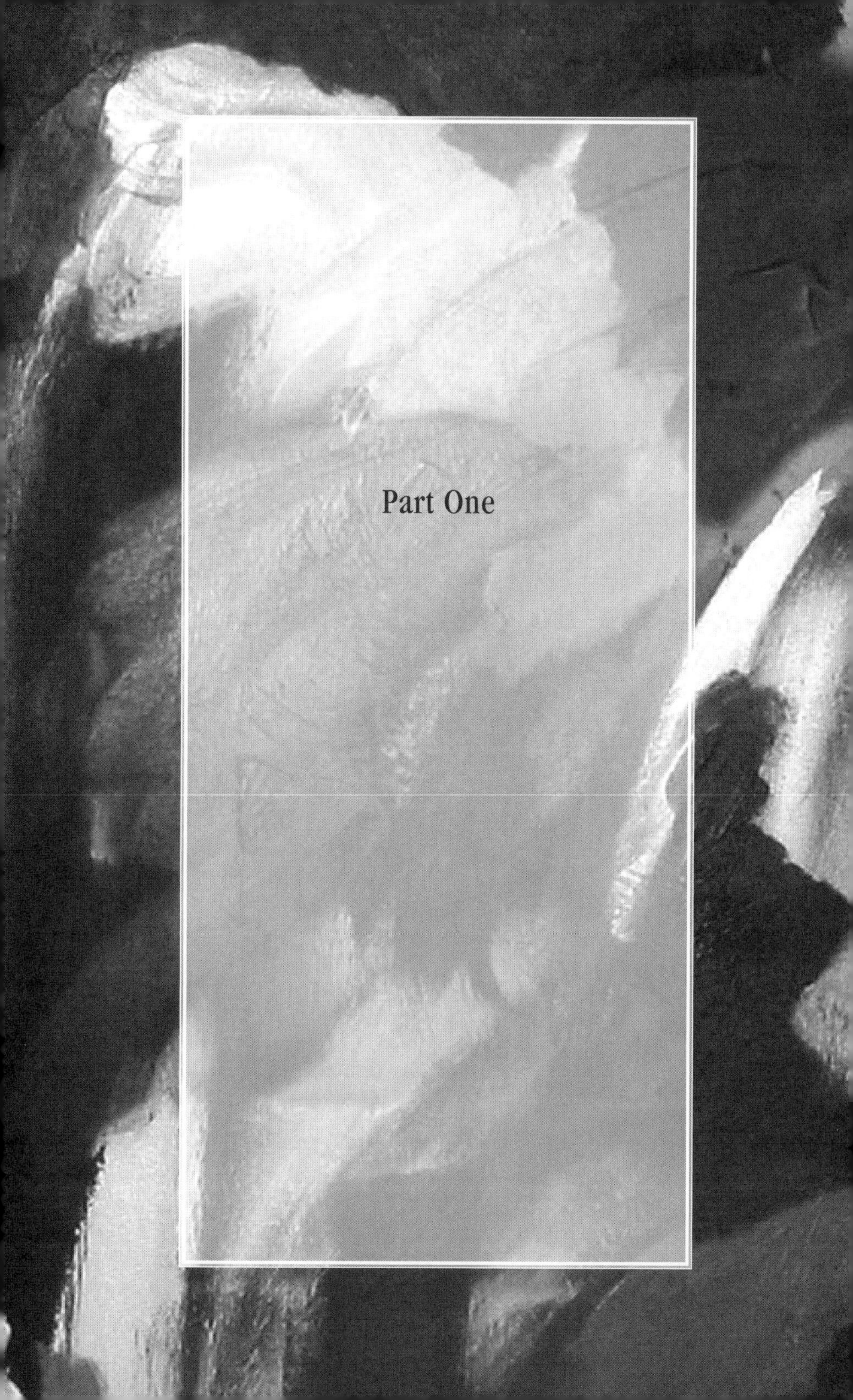

Part One

Ghosts of Spring

Each year kudzu rampages,
wielding its spear of breath,
its infallible verse,
the rattle of my elders.

In the heart of the rose,
my mother dies,
each unfurled petal
cradling in its red palm
her last muffled scream.

My father convulses
in the stamen of the iris.

The monster of May
shakes its fragile crib,
learns to walk
in the gauntlet of the dead.

I Am Not Ready to Nail This Door Shut

to Christie

For you,
I have carved a sitar from my spine,
strung it with still-sputtering veins,
composed songs in the cold, clinging darkness.

I collected acorns, planted them in the gray sidewalk.
I am inviting bulls and a lioness to sleep on my porch.

Bring whips and liniment,
steel and bamboo, thunder,
the breeze that rubs its face in the ocean's bosom.

We will bellow until oaks bend over backwards
and grass plugs its ears with marble;
dream in a bed of wisteria,
sore throats soothed by sweat and honey,
the new beets red with our blood.

Trying to Remember

I am descending the ladder,
passing through scabs
into a room where my breath lives.

There are things here with eyes
I cannot look into, my mother's brown teeth,
the way she holds up a handful of my hair
as if it were a rare pelt; her late night visits,
entrance, approach; and then, always,
the projector breaking down.

I have passed through tollbooths,
wrestled tigers as thunder went fetal;
and still, that thing in my chest
that closes like a hand squeezing a ball
yanks me back to this room
to stare into a darkness that doesn't flinch.

I hear my own footsteps on the cobblestone,
the tightrope of my sternum.
I grovel on white floor, in pools of ammonia,
collecting loose strands of hair,
tracing string on a doorknob
down to the bloody tooth.

Later, like a near death survivor
trying to recall the voice of God,
I twist the knob on the microscope,
searching for bruises on my crotch,

red blotches on my scalp, but they are not there.
They are never fucking there.
In the fretted brow of a greenhouse-morning,
I wonder, again, if they were ever there.

Well, were they?

In Praise of Us

We are the winners, you and I,
traveling dirt roads that lead to junkyards,
hurling the thermometer into the crocodile's mouth.

Always, their voices have been behind and before us,
that we have not walked by the tape measure
or turned our songs to science.

They gave us maps and exiled us,
laughed when we arrived at dry waterholes,
shook their heads as we ran with cheetahs.

But you and I,
we live in the center of the web,
feasting on our heartaches,
turning stone to water, icicles to lava,
using crosses as kindling.

We stand in the fusillade,
refusing to camouflage ourselves.
Every bullet swallowed turns to gold in our bowels.

Tradition

A pregnant woman limps in a field,
gathering cracked guitars and husks.

Shattered glass lines a floor.
Voices shrink and swell like boils.
Dogs are growling.
Fever clutches like a bur.

The father suckles his belching gun;
by day, chasing ghosts through peach orchards;
by night, a dim room
where drunkenness settles like soot.

His son follows, eyes like dull mica,
heart like an orphaned foal
starving in the wilderness.

Berlin

for Alfred

Paintings are dragged from museums,
stoned like ancient prostitutes,
books drawn and quartered like traitors.

Faces are sculpted as if from clay,
round cheeks, curved lips
blunted beneath the political knife,
eyes blasted in a kiln of blame.

Barbed wire is hung.
Smoke rises over steeples.

Sixty-odd years later,
I stand in a glass monolith
praying we have overcome,
that the crimes of the past
are like salmon that died mid journey.

But everything, or at least its legacy, returns–
most of all that envy

lodged in the human heart.

Reclamation

I held fire and ice in one hand
and witnessed neither sleeping;
walked to the swollen river,
after the rain ended,
and painted myself with mud.

Gored by the horn of the bull,
I bled on wet moss,
offered my breath to the stones.

You should have seen me,
mother, on those red hills,
singing as I tore down fences.

Wisdom, like the wind, came in gusts.

Pondering an Autobiography

I search for water
in broken hourglasses,

drive nails into the ozone
until scaffolding collapses beneath me.

Behold my entire story
in swatches and shards of china,
statuary carted off in dump trucks!

Running dry dirt through my fingers
has led me back to the same hearth,
the one that never housed a fire,

with a flue closed as tightly
as a bruised child's mouth.

This Story

I dragged it along like an interminable scab,
toilet paper stuck to the sole of my shoe.

Geometry became sacred,
the fried food
of lines and right angles.

God paced between Earth and Venus
like a homeless man, pushing a wheelbarrow
full of broken clocks. I saw the teeth of the sky.

Watching the Gyroscope

The moment is like bright light,
Jehovah's name emblazoned into my pupils.
Civilization cries, doomed messiah
left in a basket at a stranger's door.
What are we responsible for,
besides the chalkboards and sweet denial?

As false prophets pirouette on glass stages,
days settle like old men in nursing home beds.

I can only offer faces I have on reserve,
failures I have been able to keep alive,
like polio victims on an iron lung.
I cannot birth a language that signifies
tears we have been unable to shed.
I have no desire to turn vapor into rice
in order to feed the world or redeem myself.

Legacy

After the heavy-laden breakfast,
the recurrent bout with Chronic Fatigue,
I brooded while wind beat against the house,
my tongue squirming like an injured worm.

When I awoke, scrambled, still
divorced, devoid of family, dust motes
hung in the air like omens.

The sky was streaked in red,
cumulus smeared with gods' blood.

Later that night, blank darkness
loomed like a roadblock.
I thought of my committed mother,
wondered if I'd ever get hard again.

Homecoming

I am still terrified when doors are left open,
I hear a car coming up the driveway.

I am taking dictation from my body.
I am holding an auction, but
who would want these things?
A toy engine with a blood stain on one of its wheels,
a testicle resting on the bottom of a pill vial?

Every morning for years,
I have masturbated onto this sterile ground.

Now I no longer trust the moon
or consider water to be my wife,
standing here, shivering, in the leaves.

Reconciling with Stillness

It is not enough
to follow a map to familiar temples,

I need
bulldozers in my stomach,
my spine bent to its breaking point,
secrets ripped from my groin
like sequoias uprooted in a hurricane.

I am filling holes, but also
crawling into them,
refusing to suck distraction's oozing nipple,
even when my nerves
vibrate like a cheap doorstopper,

I feel loose dirt piling over me,
thirteen gravediggers burying me alive.

Sunday Night

My mind,
great miser, xenophobe,
flings open its steel door
and marches to its property's edge.

In the distance, a voice,
a drumbeat drawing closer.
The moon is an opal
sewn onto a gray-black tapestry.

The thing that kicks inside me
is neither a fetus nor a ghost,
but something else I am waiting to name.

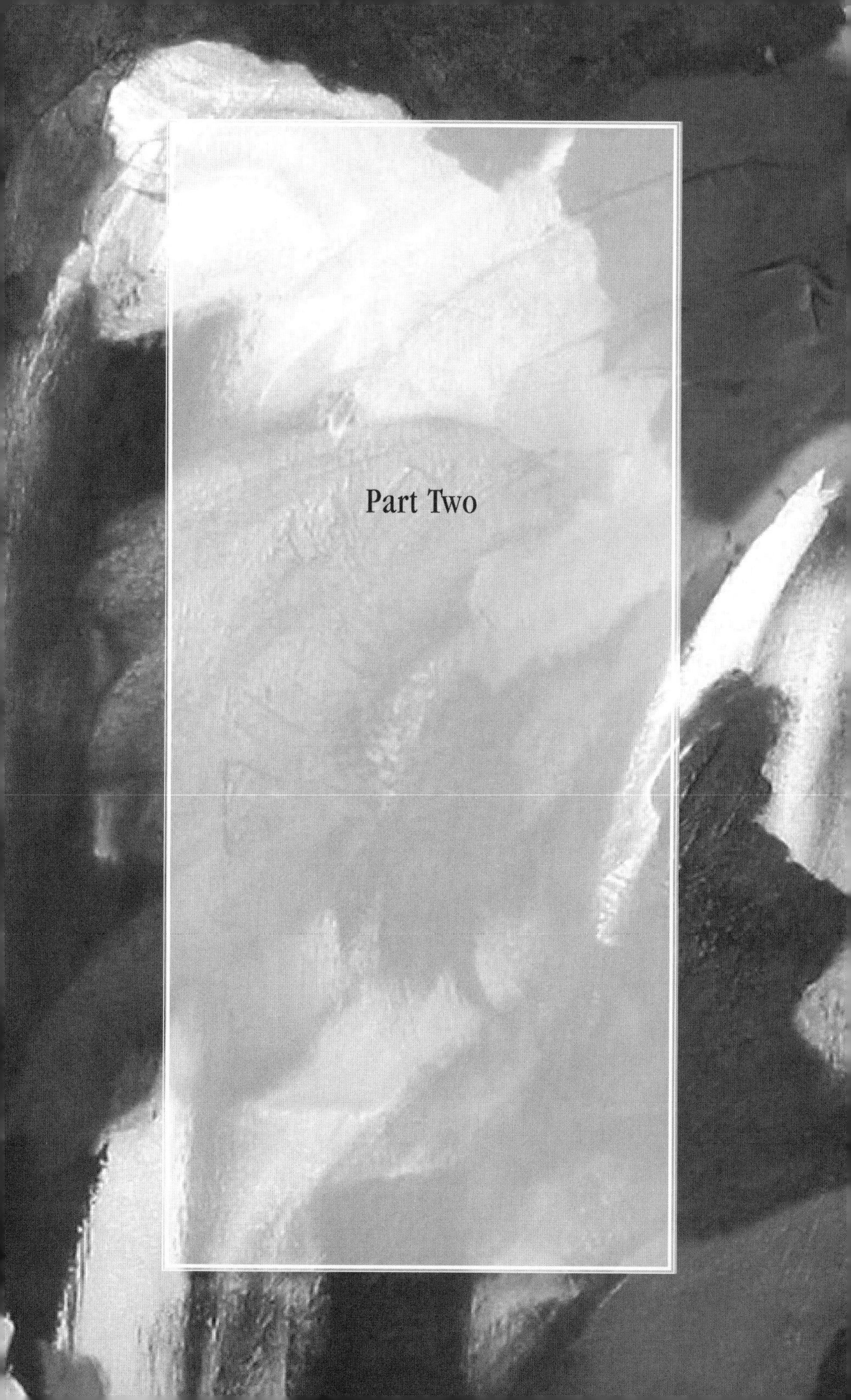

Part Two

On a Morning Such as This

Newsmen play cards inside the morgue
while Franciscan monks and Jesuit priests
flagellate themselves with rose stems.

The dream ends like a bad movie;
we arrive bankrupt at the concession stand
as night swallows like a trash compactor.

Foundations shake like a man with palsy.
Cannons are cleaned like ears,
altars scraped as if for skin samples.

A clown approaches me in the battle zone and asks,
"Why, friend, when I dust the rooms of my defeats,
do I keep finding God's fingerprints?"

Too Much to Swallow

"our nerves are whips held in the hands of time"
—Tristan Tzara, "The Approximate Man"

Darkness, falling like a cane,
bruises the day's pale limbs.

This is me on the verge of extinction,
a character in a parable
who keeps sandpaper in his pocket,
squeezing the exercise ball
until everything touched
crumbles in a death grip.

Once home, my breakdowns rise like hosts to greet me,
photos with cracked frames, brass pots collecting dust,
atheism nipping at my heels like an angry lap dog.

I light a cigarette, my first in six weeks.

Osip Mandelstam's Last Letter

Here, even fire gorged with wood burns low.
Blue flames nip at the sky like wounded dogs.

Every day we begin at the
northernmost point of endurance,
hiking like ghosts toward Kolyma;
afternoons of anemic silence,
black nights spent shivering,
listening to the screams of madmen.

Petersburg seems more beautiful now
than I thought it at the time,
spring days with Akhmatova and Nikolai,
wine as endless as a child's wishes.

Should you receive my letter, Nadezhda,
think of me, on the outskirts of Vladivostok,
snow packing like a coma, my dreams
still as dazzling as wild flowers.
Oh God, Nadezhda, the sun
is an old friend whose name I have forgotten.

Manic Summer

Every August I go mad,
overdosing on green pills,
wanting to be a hydrangea
or the satyr on the side of the birdbath.

I sprout grass from my armpits;
my navel becomes a rose garden.

Midnight drapes its arm over my neck
like a serpent dangling from a tree branch.
"You are God," it whispers.

Tonight I believe it.

Monologue for a Massacre

I dissected a drop of rain
and discovered a poultry farm,
thousands of hysterical chickens,
necks snapping to the sound
of an engine that wouldn't crank.

Later, in the psychedelic bathroom,
my house filled with drunkards,
I pried open my skull to witness
black children in chains
running naked through rows
of burning apple trees.

In my best moments,
I still come to with stomach cramps,
heart thrashing like an epileptic in a straitjacket;
above me, the billboard of a fading rainbow.

I will never be able to defecate all the rage I have eaten
or make peace with elephants lying mutilated in the brush,
ivory tusks hacked into cheap gewgaws,
much as we, too, have been hacked, into men.

After the War

"The day of remembering comes again."
—Anna Akhmatova, "Requiem"

I adorn myself with Jewish symbols,
paint my scrotum green,
drag the dull blade across my torso.

I am mixing ink
from phlegm, feces, blood.
This is what I will use
to write my manifesto.

The ones who went before me,
who tumbled into pits, are waiting.
They stand in a meadow,
staring into red sky.
Behind them is a great city;
a place, perhaps, called heaven.

It is burning, and everyone knows
God is the one who started the fire.

Original Sin

Albino bulls groan in a bloody field
as deacons bury the cheerleader's ruptured hymen.

Look at the buffet, the fish and wine.
Who arranged this table? Who invited me here,

forgotten farmyards like stale wafers,
lepers singing in a monotone, the anxiety

of not belonging grinding like a disposal?
Black holes swallow stars in a forced ouroborous,

as if fate itself were plagued by heartburn–
as if I could shatter these chains and know

immediately how to stand, in what direction to walk.

Apollinaire's Confession

You who speak to me of trivial distinctions,
I am bleeding in a barren field,
counting my dead by night,
and still I insist, nothing happens to a man
he doesn't secretly wish for:

Drought spreads like a virus
through my body's orchards.
I pick fruit that hangs withered
from boughs as lifeless as eyelids on a corpse.

When the sky turns pallid, though,
and clouds fill like a heart
needing to divest itself, I banish
the possibility of rain, clutch
desiccation as if it were a diamond,
holding tragedy close to my chest
like a mother who cannot bear the idea
of her child becoming a man.

In the Imbroglio

I carry welts on my hands
to prove I have spent
my share of nights
strapped to the electric chair.
It is all I can do to sing among the dead.

The rain arrives like a missionary,
a conversion that spares the nerves.
I harvest fruit in sight of vultures,
celebrate by gargling dust. We are each
allotted only so many awakenings.

Each day I digest something I have not yet eaten.
I am used to that, acids bubbling,
trains roaring in my rib cage;
my stomach turning like a press,
trying to give voice to something not there.

The Connection

The day is as stubborn as an atheist,
moments running like mascara on a prostitute's face.
I hold my doubts as if they were a crystal ball;
sleep in odd places, fancying myself a locust.

The only flag I ever waved was my own tongue
after my father yanked it from my mouth
as if it were the dipstick in his '62 Thunderbird.
I am glad it is raining, that my family secrets
are finally being flushed from this crawlspace.

For years, I played the fool in my own court.
The tower I lived in was made of black onyx.
I clutched the lie as if it were a charm,
psyche split like a pierced abdomen;
my own breath was my mother's period.

Here in Sodom

A witch doctor raps in monosyllables
while the flame heats the spoon
and a disgraced soprano shatters glass.

In fallout shelters, we celebrate
abnegations and abdications;
each day we go bankrupt and horde a fortune.

Light has to be imported;
death is exiled; in the metal sky,
gods lean forward, shattered but curious.

What Is There But Crescendo

Dolphins swim from my mouth,
sharks from my groin.
The house trembles like a pontoon.

My breath passes through my loins
like frayed fishing line, emerging from my
nostrils like a hook, while the marlin

of those things kept from myself
continues to plunge.

What Things Mean

Dogwood blossoms assure me
I will see the ocean again,

always be a mystery to myself,
my desire a carrier pigeon still learning to fly.

The rain slices away strings
that bind my tears and penis.

I walk the streets like a holy cow,
remembering the earth is not a chessboard,

even if everything else is.

The Ontology of Dying

"And round about his home the glory
That blushed and bloomed
Is but a dim-remembered story ..."
—Edgar Allan Poe, "The Fall of the House of Usher"

Daily the blueprint waxed more labyrinthine;
convoys of trucks wound through the ravaged plot,
delivering colossal spools of wire, miles of virgin pipe,
rare timber from Amazonian forests. Expense
grew like a tumor. Electricians, plumbers,
carpenters scurried like slaves in a concentration camp.

I let them build it, watched like some ancient Greek
crashing the chamber of The Fates,
observing his own destiny being woven.
Board by board, floor by floor, like a scientist
creating a clone, they erected the house, despite the moon's demurral.

For weeks, I wandered its autogamous maze of corridors,
jamming empty rooms with gewgaws, ottomans,
divans, plush wardrobes, and impenetrable tomes.
By diurnal light, I loitered in the garden, watched the landscape
flex its muscles like a rabid dog stretching a chain.

The boxwoods grew in perfect orbs, antiquating the gardener's shears.
Roses bloomed, untouched by worm or beetle,
standing in perfect rows like Nazis at attention.
Azaleas, camellias, and gardenias blossomed in perfect proportion,
emitting ambrosial redolence, immune to the temptation of overgrowth,
requiring neither pruning nor cultivation: the apotheosis of purity.

Tourists came by the busload to gaze upon the architectural
wonder; aspiring poets paced in rapture, scribbling dithyrambs
with Dionysian facility. I conducted tours, flinging wide
doors to my most private chambers. Engrossed strangers
ogled my treasures and trinkets with a smutty awe.
Winter never came.

The house, as sterile as an operating room,
sanitizing its own guts like a self-cleaning oven,
required no maintenance. The sentinel of the sun, never questioning
orders, yielded its watch to the night without the protestation of dusk.
Midnight felt as safe as a bed, lingered as quiet as a convent.
The teeming Eden, with its inexhaustible prowess, preyed upon me like a Minotaur.
As I grew pale and my appetite waned, wisteria bloomed more fervently.
The walls of the mansion shone like an infant's scrubbed cheeks.

I flipped through thousands of pages of unintelligible works
like a philologist trying to crack a code; I hurled vases,
mirrors, cups, plates, and grails upon a floor of stone.
They were unbreakable. I rushed into the courtyard
armed with a chainsaw, the gnawing teeth soon
ground to silence by bronze stems and steel limbs;
cracked the skull of an ax on a trunk made of iron.
I wept like a prisoner in a soundproofed cell,
screamed out in the night like an atheist in a riptide.

One morning, my arms and legs finally shed like withered petals,
my body sculpted into reptilian litheness by the chisels of time and necessity,
I slithered down marble steps, across plush carpets,
through a malaise of rooms into the garden,
where the maniacal sun greeted me like a spotlight on a guard tower.
As I crossed the boundary of the estate, like an inmate cutting through a fence,
I heard the crowd, behind me, pounding on the golden door.
I didn't look back. I crawled on my cavernous belly until I reached
a nameless place where clouds eclipsed the sun, where I was soaked by rain;
where autumn arrived and green leaves browned; the restless sap
swallowed its song; the creek's throat froze over; the night was filled with ghosts.

Like an exile, I set up camp there, burrowing into the earth,
subsisting on rodents, berries, husks. Seasons arrived and
departed like gurus. I evolved toward a certain complacence,
like that of a flash of lightning content to foil
the monologue of thunder.

A Calling

I emerge,
the black widow on my shoulder.

Where is my father?
Where is someone who will walk
through the river's bed barefoot?
Hold a hornet in the palm of his hand?

In the valley, horses stampede into barbed wire.
My tongue has been turned to paper
upon which a great lie is written.

We must travel together
to where the compass goes blind,
you with your rags, I with my bag of bones;
when night forms like a scab,
lie in brambles singing cold, sober songs,
memorizing the faces of our fear.

I will celebrate your story.

You must celebrate mine.

Part Three

Hiding

for Richard

I spend the morning
looking at photographs of my dead sister,
dark mannequin posing
beside husbands, parents, siblings,
her son–people who look like extras on a movie set–
the years' battering superimposed on her face,
reminding me of Holocaust images, olive-skinned
girls who died in showers at Auschwitz.
Even in the photo where she
wades in a nurturing Atlantic, she
reminds me of some Jewish Ophelia, her
moribund drama hemorrhaging into the spindrift,
thick shadow snuffing a nirvanic beach.

Last night a friend told me she felt
my ex-wife had not been good for me,
that I had hidden behind her like an eclipsed sun,
and I thought about how my own mother was a piranha
who each morning at the breakfast table
stripped her sons and daughter to the bones.
Years later, my father would tell me
he sacrificed his children to appease his wife,
offered us to her as if she were some pagan goddess
who needed to drink daily her own family's blood.

We all learned to hide; it is our legacy–
my sister and I, even my brother,
skulking in the custody of his own rage.
We grew out of childhood
like houseplants in a hurricane,

domestic pets abandoned in a jungle;
floating out of body in public places;
passing like ghosts through marriages and jobs;
watching ourselves fuck spouses and greedy strangers,
naked bodies move; not recognizing ourselves, honestly
not knowing how we were going to survive the relentless invasions,
the ambushes and slow, secret military movements,

this thing other people simply called life.

The House I Grew Up in Was Never Mine

Satyrs wandered the copses looking for children;
across the county line, in the church,
spinsters flossed with verse,
gave immaculate birth to rapists.

The alarm that shattered the morning fog
shone on a makeshift altar while my grandfather
sharpened his cuspids and pretended to read.

I reached down my gullet a thousand times,
but could only graze the serpent's tail.

I was surrounded by sand and broken crucifixes,
whether I dove into my own hemophilic blood
or prayed to the planets, spinning above me like severed testicles.

Towing the Debacle

I spend nights combing rest areas like a
mad swan seeking its mate. I
am weathering awkward silences,
accusations like glass bottles
thrown from a moving car, walking
tightropes with a mastery
reserved for the insane.

I watch my towers,
meticulously crafted complexes,
crack in half like a wishbone.
These bridges, burning like tapers,
never led to a place where gold
was for the taking. Corn grew
on the other side of the razor wire,
but I was never able to see God
in beautiful things. That
was my shortcoming, one of them.

The Lonely Vacation

Empty chairs melt under a blue umbrella.
A toothless man, standing in water
ankle-deep, holds up a dead shark.
His son, bleeding from the eyes,
flogs himself with seaweed.

A naked woman smiles at me,
breasts heaving like buoys,
tar trickling from her thighs.
Her husband, chewing a fishhook,
is dressed in tinfoil,
wedding ring burning like a nova.

The lifeguard recites sonnets
to a teenage girl wearing headphones.
Tourists flock down boardwalks
to see a messiah turn chicken bones into IOUs.

Floundering

The bare tree rooted in dry earth
speaks to the world of my bleeding groin.

I still carry those days as if they were medicine balls,
hungry hands in private places.

It is a book that never ends,
a magic eye eternally out of focus.

Crows perch on my gray branches,
vultures gnaw on my leaves.

Whose criminal name should I attach to these visions?
I have been poking in a grave all my years

and have still not found the infamous corpse,
a sack of bones, even a ghost

who could point me in the right direction.

The Bamboo Shoot

Does it wonder what thick patch it hails from,
what vast acreage of verdancy it could call home,
meanwhile peering from this hollow fort
like some sheared Rapunzel,
aging in gray water, waiting for small,
slimy accomplishments to emerge?

When I look at myself, I see the places
where axes severed me from my many mothers,
how I have drifted far from the places where I was born.
Now I look for links as invisible as electricity,
comminglings that rise in the gut like phantom pain.

I see how this bamboo shoot spends its days
mumbling private mantras while fine roots
press against glass like children trapped in a flooded house:
Everything needs to blossom; at the same time,
feel the pulse of something else growing.
Let me crack through the glass of my own thick vase.
Let my hungry roots find you, my green life
merge with yours, even if it can't be seen.

Fear of Death

for Chris

I remember, as a teenager, laboring in the
peach orchard, but at day's finish, when the sun
hung over the mountain like a drooping eyelid,
abandoning ladders and pitchforks,
throwing down baskets as if I were God
and it was the dawning of the seventh day.

Then, things began and concluded;
days were born and demised; red earth
taught me the aesthetics of death.
But when I came to the city,
a restless soldier in need of war,
I began murdering myself for a living.
Like a child in the fifties taught to hate Russians,
I grew to despise death. I locked my thoughts
and bulked my body against its entry.
Days stretched like a repeating decimal.
Life became as irresolvable as pi.

It becomes a habit, to never let a
thing die– devouring caffeine pills,
making love while playing chess on a cell phone,
terrified to yield to that persistent Satan, sleep.
No one can be still long enough to crawl into a cocoon,
why else are there so few butterflies in our midst?

Where Do We Go from Here

I pretended that spring was my sister,
summer my concubine,
that my ambitions were blessed by the sun.

I beat my effigies as if they were piñatas,
finding nothing inside them
but dry bones and the stench of formaldehyde.

What will we do now,
watching grass grow
like stubble on a rapist's face,
knowing the altars we destroyed
were never holy?

Where do we go from here,
now that words are lost
like baubles in an earthquake
and silence swarms around us
like the vultures of an empty prayer?

I Certainly Am No Metronome

Tumbling into a cracked kaleidoscope
like an adolescent riding a waterslide–
motels and fast-food restaurants,
waitresses whose names blur
like a face seen through fire.

The bricks in the middle-class house
are breathing; the loins of the sky heave;
the cosmos holds its breath like a gambler
as presses chortle and cough,
kindergarten teachers dream of high-rises,
and the fanfare of trendy addiction subsides:

Now we lean on ancient memories as if they were a cane
and teach our children death is a virus.

Disruption

"Is not the story of Adam summed up in me?"
—The Infancy Gospel of James

Single moment
like a twitching eye.
Broken wing
flapping against the sky's pale hand.

The cocoon grows tired
of the miracle it houses.
Fire vomits oxygen.
See the blade of grass
in the morning,
hands outstretched
to keep the dew at bay.

What is this now,
that even light
looks over its shoulder
as it races through darkness?

At This Hallowed Moment

Angels lurk behind colonnades,
haloes dying like an untended campfire.
Like a dead bird, silence drops–the tired sun,
neurotic minstrel, croons its madrigal
in the key of frustration.

I am a master at building temples
in which I never worship.
Desire becomes my Trojan Horse.
Morning arrives like a cop delivering a subpoena.

Standing like a leper-king beneath a barren fig tree,
I am ready now to garb my quest in sackcloth,
to drive nails through the palms of everything I know.

Christening the Dancer

Ash flutters like a butterfly beneath a bell jar.
My darkness is as thick as grease.

"Fine, now tell me something of love,"
the dancer inside me screams.

The wind wraps like a bandage
around my blistered feet.

The dancer is clawing through my lungs,
clawing through my back,
bursting from my flesh
like a channeled text, leaving
splinters of bone on the floor.

"Is it my turn now?" the dancer says.

"Yes," I say, "now it is your turn."

The orphan dies in an empty room.
The creator dies on the seventh day.

"Look," the dancer says, *"I am flying."*

The great gleaming eye above the clouds
bears down like a smoldering drill.
Stars are bursting like bubble wrap.
Scaffolding is removed from my genitals.
Thermometers lie broken on the surface of a thawing lake.

I am becoming the dancer,
the dancer is becoming me.
I am becoming nothing,

I am becoming choreography.
I am as empty as tomorrow's urn,
and the dancer is dancing
on the stage of my emptiness.

And the dancer is dancing
as blood pours from his palms
and water spills from the hole in his side.

"I forgive you," the dancer says.

"I forgive you," I reply.

The song spreads like a frayed seam.
The dancer bleeds dust, the dancer
drinks light. Like an old god,
the dancer dies and is reborn.

And his name, whatever it is, is not mine.

Secrets

How many childhoods have come and gone
like lassoes, how many incarnations
spent picking with broken fingers
at that Gordian knot called sin?

One thing I am sure of,
even if the Sphinx's riddle is solved,
I will still be devoured.

The Rapture will come and go
like a snow flurry in the deep south.
What will remain are the things
we have built highways around,
pupae writhing inside steel cocoons,
all we have learned not to speak of
when the sun goes down.

The Reawakening

to a father

"What is this sleep which holds you now?
You are lost in the dark and cannot hear me."
—Epic of Gilgamesh

The hand in the pocket is a fist.
Even the white moon
conceals a jealous heart.

You taught this,
gave me the boa of skepticism,
advised me to study suffering,
scorn the sufferer.

Like an initiate in a cult,
I wandered rivers and swamps
feeding on stalks and algae
until I reached a frozen sea
where I sat as prescribed,
the naked lotus.

My reptilian legacy
wound about my torso
like a steel girdle.
Beneath our weight,
ice cracked its vow.
We plunged together
into a rimy uterus.

In that dark womb,

the coils grew flaccid.
Sap surged through my veins
like electricity restored
after a power outage.

I threw the dead thing from me
as if it were a kudzu vine
yanked from a sapling.

I saw far above,
like a star beyond
a black hole's suck,
a pinhead of light,
my arm extended toward it
like the bow of a ship.

Gliding up the canal
toward that shimmering orifice–
a root bursting its husk–
I crowned the surface.

Unfamiliar tongues conducted me,
my gasps harmonizing with the dawn.
As if your spell had been broken,
my pores blossomed arias.

Like an amnesiac suddenly remembering,
I recognized the palm trees,
that I was not perishing in some boreal sea,
your constrictor crushing my rib cage,

but lying on the breast of a warm beach
as if in the arms of a wet nurse.

Surrounded by relieved faces, I saw
open hands– they resembled
my own.

After the Exhaustion

I tore down entire cities
to study what lies beneath
concrete slabs, tilted foundations,
the palimpsest of epitaphs.
Grief is leading me down
a dirt road of madness
toward an abandoned town called sanity.
My story is ripening like tomatoes in August.
How can a man pass through turnstiles and tollbooths,
put his signature to a thousand daily contracts,
and yet fail to learn his own name?
What happens in the factories,
the cafes and salons, is as holy as what
gestates in the dusty sanctum sanctorum.
These are two worlds, but they are like the
moon and the sea, and I am living in both.

Acknowledgements

The author wishes to thank the editors of the following journals, in which these poems first appeared:

Adirondack Review *(www.adirondackreview.homestead.com):* "Christening the Dancer," "Floundering"

Blue Fifth Review *(www.angelfire.com/zine/bluefifth):* "Disruption"

Branches Quarterly *(www.branchesquarterly.com):* "Apollinaire's Confession," "Tradition," "What Is There but Crescendo"

Disquieting Muses *(www.disquietingmuses.com):* "I Am Not Ready to Nail This Door Shut"

The Drunken Boat *(www.thedrunkenboat.com):* "The Reawakening"

Ludlow Press Journal *(www.ludlowpress.com):* "Berlin," "In Praise of Us," "Sunday Night"

The Melic Review *(www.melicreview.com):* "In the Imbroglio," "The Ontology of Dying"

Pierian Springs *(www.pieriansprings.net):* "I Certainly Am No Metronome," "Original Sin," "Watching the Gyroscope"

Poetrybay *(www.poetrybay.com):* "A Calling"

Poetry Repair Shop *(www.poetryrepairs.com):* "The Lonely Vacation," "On a Morning Such as This"

Red River Review *(www.redriverreview.com):* "After the War," "Too Much to Swallow"

Samsara Quarterly *(www.samsaraquarterly.net):* "The Bamboo Shoot," "Hiding"

Sidereality (www.sidereality.com): "Manic Summer"

Small Spiral Notebook (www.smallspiralnotebook.com): "Osip Mandelstam's Last Letter"

Three Candles (www.threecandles.org): "After the Exhaustion," "Fear of Death," "Homecoming," "Monologue for a Massacre," "Pondering an Autobiography," "Secrets"

Thunder Sandwich (www.thundersandwich.com): "Here in Sodom," "This Story," "Trying to Remember"

2River View (www.2river.org): "At This Hallowed Moment," "Ghosts of Spring," "Reclamation," "Where Do We Go from Here"

John Amen received his bachelor's and master's degrees from the University of North Carolina at Charlotte. He has published poetry and fiction in various magazines and journals and was recently nominated for a Pushcart Prize.

He has traveled extensively as a performing musician, both with a band and as a solo act, and has released three full-length recordings. His fourth recording will be released in early 2003.

He is also an artist, working primarily with acrylics on canvas. Further information is available on his website: *www.johnamen.com.*

In addition to his literary, musical, and artistic endeavors, Amen founded and continues to edit the online literary bimonthly, *The Pedestal Magazine* (***www.thepedestalmagazine.com***).

He has lived in New Orleans and New York, and currently resides in Charlotte, North Carolina.